YOUR KNOWLEDGE HAS VALUE

Testing statistical hypotheses using parametric tests

Stonia Thorand

Bibliographic information published by the German National Library:

The German National Library lists this publication in the National Bibliography; detailed bibliographic data are available on the Internet at http://dnb.dnb.de.

ISBN: 9783346668271
This book is also available as an ebook.

© GRIN Publishing GmbH
Trappentreustraße 1
80339 München

Print and binding: Books on Demand GmbH, Norderstedt, Germany
Printed on acid-free paper from responsible sources.

The present work has been carefully prepared. Nevertheless, authors and publishers do not incur liability for the correctness of information, notes, links and advice as well as any printing errors.

GRIN web shop: https://www.grin.com/document/1243694

2022

Testing statistical hypotheses using parametric tests

CASE STUDY

STONIA THORAND

B.SC DATA SCIENCE

16.05.2022

Table of content

List of figures

List of tables

List of Abbreviations

Abbreviation	Description
MS	Mean square
SS	Sum of square
df	Degree of freedom

1. Introduction

The pandemic has spread during the last two years dramatically. In Germany alone roughly 25,66 million confirmed cases and 137 348 confirmed Corona deaths were documented (World Health Organization, 2020). Interestingly the highest number of confirmed cases was around 1,6 million on the 21st March of 2022 while the confirmed death counts were at 1 520 in Germany at the same day. Conversely, this means that the counted deaths are roughly 1 000 times lower than the infection levels, while the highest death count was detected at the 14th of December in 2020 with a value of 6 410 with roughly 170 00 confirmed cases (see Figure 1).

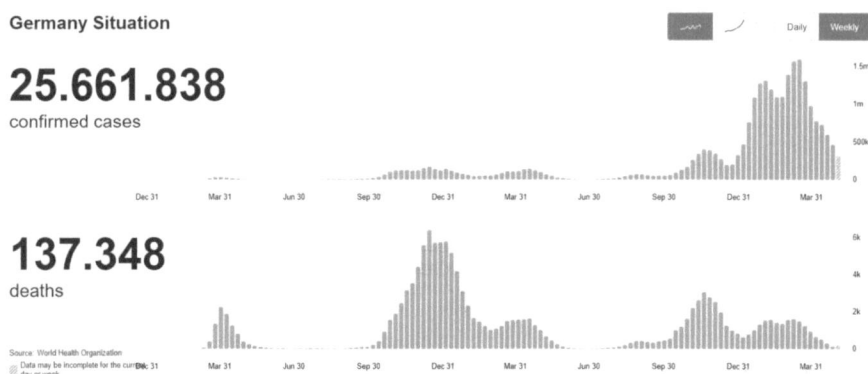

Figure 1: Chart of confirmed cases and deaths of Corona patients in Germany from December 2020 to May 2022 (World Health Organization, 2020).

Meaning that around 27 times less deaths were confirmed in comparison to infection cases. The decrease of Corona deaths may be caused by the introduction of the Corona vaccinations. With the help of the accumulated data during the pandemic, the effect of the vaccination can be tested against the hopitalization values of Corona patients with statistical tests (Robert-Koch Institut, 2022). The test does indirectly adress the issue towards the decreasing Corona deaths with increasing infection cases. The increasing value of confirmed cases clearly shows, that the vaccination does not inhibit the infection, but somehow may have an weakening effect on the course of the disease. The approach mainly focus on different vaccination status and age groups of hospitalized Corona patients in Germany. Not included are factors like the Corona variation and the specific vaccines. Statistical tests to address such questions are parametric tests, which are powerful tools to evaluate data following a normal distribution. Within the assignment, two parametric tests are presented (pearson correlation coefficient and one-way ANOVA), while the one-way ANOVA was chosen to adress the question of interest desribed above. With the data extracted from the Robert-Koch Institut (Robert-Koch Institut, 2022), an analysis was carried out to determine significant effects of the vaccination status of Corona patients on their hospitalization. All calculations done in the assignment were generated via Jupyter Notebook (Thomas et al., 2016). The results of the analysis indicate significant differences between hopsitalized Corona patients with different vaccination status, which is discussed in detail in the conclusion.

2. Parametric tests

Parametric statistics assumes that the sample data follows a normal distribution. The term parametric statistics was first mentioned by R.A. Fisher in 1925. Parametric tests are widely used, leading to the development of multiple methods to answer scientific questions within the parametric statistics, some of the examples are:

- ANOVA tests, which measures if differences between two or more groups are significant
- Pearson`s correlation coefficient, which measures the linear association between two groups

These two examples will be further explained in detail in the following sections.

2.1. Pearson correlation coefficient

The Pearson correlation coefficient measures the degree of relationship between two linearly related variables. The measure is obtained by the following formula:

$$r_{xy} = \frac{n\Sigma x_i y_i - \Sigma x_i \Sigma y_i}{\sqrt{n\Sigma x_i{}^2 - (\Sigma x_i)^2} \sqrt{n\Sigma y_i{}^2 - (\Sigma y_i)^2}} \tag{1}$$

With following definitions:

- r_{xy} = Pearson r correlation coefficient between x and y
- n = number of samples
- x_i = value of x for ith observation
- y_i = value of y for ith observation

In order to obtain the Pearson's correlation coefficient, some requirements have to be met. First, this method does belong to the parametric tests, which means, that the values are expected to be normally distributed. Also, the variables need to be linearly related, meaning that between the two variables there should be a straight-line relationship. After the implementation of the Pearson correlation formular following results can be determined (see Table 1).

Table 1: Strength of the correlation determined from the result of the Pearson correlation test.

	Pearson correlation coefficient, r	
Strength of correlation	Positive	Negative
Small	0.1 – 0.3	- 0.1 to -0.3
Medium	0.3 – 0.5	-0.3 to -0.5
Large	0.5 – 1.0	-0.5 to -1.0

Correlation in the context of Pearson's coefficient explains how one variable differ, when the other variable changes. If the result equals one, this means that one variable change in the same extent as the other variable. The dependency decreases with lower results up to 0.1. This result indicates that the change of one variable is not significant to explain the change of the other variable, since they don't correlate. Regarding the positive and negative results, this indicates the direction of the relationship between the two variables of interest. Positive results indicate that the increase of one variable leads to the increase of the other variable. The negative result with a large correlation for example would mean, that the increase of one variable would lead to the decrease of the other variable in the same extent.

Despite the powerful statement generated from this parametric test, also another method is discussed in the following chapter.

2.2. Analysis of variance (one-way ANOVA)

The one-way ANOVA addresses the difference between various groups. The main characteristic of this method is that it does include one independent variable and one quantitative dependent variable. Within the independent variable should be at least three levels. The ANOVA test is a sub class of the T-test. The basic idea of ANOVA is to compare different groups based on their mean. With this approach not the mean directly is compared between the different groups, here various proportions of variance are used to determine the balance between the means. The result of the ANOVA approach is the F-statistics. The F-statistics is a statistical approach, to test whether there is a difference between two or more groups. The F-statistics belong to the group a hypothesis tests. The hypothesis test first determines the null hypothesis, which does state that the mean of the groups equals the same. The goal is to determine if the hull hypothesis is true or false. The alternative hypothesis means that minimum one group shows significant difference from the dependent variable. The mathematical basis of the F-statistics is to determine the F-value. The F-value is calculated according to following formula:

$$F = \frac{MS_{between}}{MS_{within}} \tag{2}$$

Most important value for the F-value calculation are the mean square between ($MS_{between}$) and the mean square within (MS_{within}). Before implementing equation (2, the mean squares need to be calculated with following formulas:

$$MS_{between} = \frac{SS_{between}}{df_{between}} \tag{3}$$

$$MS_{within} = \frac{SS_{within}}{df_{within}} \tag{4}$$

$$df_{between} = (k - 1) \qquad (5)$$

$$df_{within} = (N - k) \qquad (6)$$

Where following definitions apply:

- SS_{within} = Sum of square within groups
  - $SS_{within} = \Sigma(x_1 - \bar{x}_1)^2 + (x_2 - \bar{x}_2)^2 + \cdots + (x_k - \bar{x}_k)^2$ $\qquad (7)$
  - k = total number of independent variables
  - $\bar{x}$ = group means
  - X = value
- $SS_{between}$ = Sum of squares for area between groups
  - $SS_{between} = \Sigma(\bar{x}_1 - \bar{\bar{x}})^2 + (\bar{x}_2 - \bar{\bar{x}})^2 + \cdots + (\bar{x}_k - \bar{\bar{x}})^2$ $\qquad (8)$
  - $\bar{\bar{x}}$ = great mean, the total mean of all values of all groups
  - k = total number of independent variables
  - $\bar{x}$ = group means
- $df_{between}$ = Degrees of freedom between groups
  - k = total number of independent variables
- df_{within} = Degrees of freedom within groups
  - N = total number of values of all variables or groups
  - k = total number of independent variables

The calculated F-value from equation (2 is checked for its significance by consolidating the f-distribution table. The tables should be picked according to the alpha (α) value which was picked prior the calculation. Alpha is a threshold to judge whether the test statistics is statistically significance. After a specific α value was chosen, the critical F-value is determined from the F-distribution by checking the value regarding the degree of freedom. The F-table has a slide for the degree of freedom denominator (df_{within}) and nominator ($df_{between}$). When the calculated F-value is less than the critical F-value from the distribution, than the rejection of the null hypothesis failed. Meaning that there is no significant difference between the tested groups. Another approach to determine the significance can be obtained via the p-value. This value shows the probability to generate results as the average of the result means. Meaning if the calculated p-value is lower than the significance level of 0.05 the probability of the null hypothesis is very low, and it can be rejected. In the following paragraph this approach is chosen to address dependency of the vaccination status of hospitalized Corona patients.

3. Real-life problem addressed by ANOVA

The extraordinary situation caused by the pandemic has been with us for two years now. Within this time, humans have been able to adapt through newly developed vaccines. During this time, data has been collected, which gives the opportunity to analyze what has happened. The main focus of this assignment is whether the age or the vaccination status are more affecting the hospitalization rate of COVID-Patients in Germany (Robert-Koch Institut, 2022). To address this scientific problem, the one-way ANOVA was taken into consideration. Here only data was considered which included patients with booster vaccination, which would start September 2021 (calendar week 42) until March 2022 (calendar week 15). Tow data sets were tested, one dataset includes the hospitalization of corona patients with different vaccination status from the age of 18-59 years (see Figure 2).

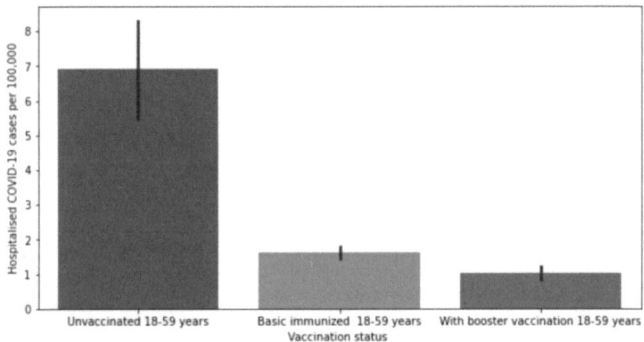

Figure 2: Representation of hospitalized COVID-19 cases per 100 000 against different vaccination status at the age of 18-59 years (Robert-Koch Institut, 2022).

The second data set includes the hospitalization of corona patients with different vaccination status at the age of 60 plus years (see Figure 3).

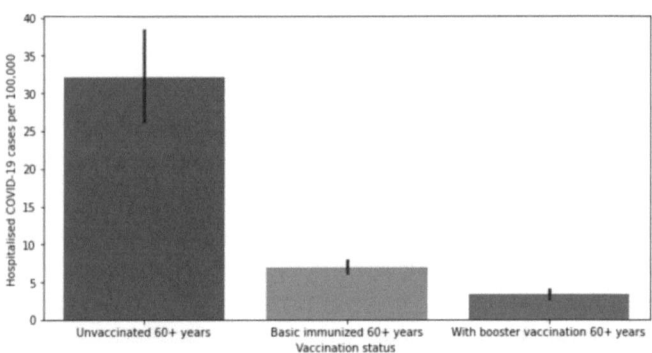

Figure 3: Representation of hospitalized COVID-19 cases per 100 000 against different vaccination status at the age of 60 plus years (Robert-Koch Institut, 2022).

To analyze the data, the one-way ANOVA was carried out via Jupyter notebook (Thomas et al., 2016), before using any predefined packages the mathematical background of all calculations needed were coded from scratch. The result was compared to the results generated from the statsmodel package (Seabold & Perktold, 2010).

The data used for the analysis can be seen in Supplementary table 1. The data consists of one independent variables with three sub-groups. The group of unvaccinated corona patient at the age of 18-59 years, the second groups consist of hospitalized corona patients with basic immunization at the age of 18-59 years and lastly the groups of hospitalized corona patients with booster vaccination at the age of 18-59 years. The goal of the one-way ANOVA is to determine if there is a significant difference in the hospitalization value of Corona patients at the age of 18-59 years with different vaccination status. To calculate the one-way ANOVA, the F-value is a very important variable, which is calculated according to equation (2). To calculate the F-value, other variables are needed and calculated according to the equations (2) to (8). First the mean of each group and the total mean were calculated as seen in Figure 4.

```
#calculate groups means
df_un_mean = df['Unvaccinated 18-59 years'].mean()
print(df_un_mean)
df_grund_mean = df['Basic immunized  18-59 years'].mean()
print(df_grund_mean)
df_auf_mean = df['With booster vaccination 18-59 years'].mean()
print(df_auf_mean)

6.903575965991387
1.629995556978079
1.0183472896997745

#calculate total mean
df_total_mean = (df_un_mean + df_grund_mean + df_auf_mean)/3
print(df_total_mean)
df['df_total_mean'] = df_total_mean

3.1839729375564136
```

Figure 4: Code for calculation of the group means and total mean of Corona patient hospitalization values (Robert-Koch Institut, 2022).

Followed by the calculation of the sum of square within and between for the data set as seen in Figure 5. The calculations were obtained according to the equations (7) and (8).

```
#compute sum of square within
df['df_un_mean'] = df_un_mean
df['df_grund_mean'] = df_grund_mean
df['df_auf_mean'] = df_auf_mean
ss_within = (sum(((df['Unvaccinated 18-59 years'] - df['df_un_mean'])**2))
+ sum(((df['Basic immunized  18-59 years'] - df['df_grund_mean'])**2))
+sum(((df['With booster vaccination 18-59 years'] - df['df_auf_mean'])**2)))
print(ss_within)


396.1141339647379


# calculate sum of square between

ss_between = (sum((df['df_un_mean'] - df['df_total_mean'])**2)
             + sum((df['df_grund_mean'] - df['df_total_mean'])**2)
             + sum((df['df_auf_mean'] - df['df_total_mean'])**2))
ss_between


544.4458977138015
```

Figure 5: Code for the calculation of the sum of square within and between of Corona patient hospitalization values (Robert-Koch Institut, 2022).

With the help of the sum of squares the mean squares can be calculated according to the equations (3), (4), (5) and (6). The code for the data set can be seen in Figure 6.

```
#calculate mean square within
number_samples = (df['Unvaccinated 18-59 years'].count(),
                  df['Basic immunized  18-59 years'].
                  count(),df['With booster vaccination 18-59 years'].count())
print(number_samples)
df_within = sum(number_samples) - 3
print(df_within)
MS_within = ss_within / df_within
print(MS_within)

(26, 26, 26)
75
5.281521786196506
```

Figure 6: Code for the calculation of the mean square within and between of Corona patient hospitalization values (Robert-Koch Institut, 2022).

The generated parameters are necessary to be able to calculate the important F-value and p-value, which are needed for the evaluation of the result (see Figure 7).

```
#compute F-Value
f = MS_between / MS_within
f

51.54252124233732

#compute p-value
import scipy.stats
p_value = 1 - scipy.stats.f.cdf(f, df_between, df_within)
p_value

8.215650382226158e-15
```

Figure 7: Code for the calculation of the F- and p-value of Corona patient hospitalization values (Robert-Koch Institut, 2022).

A shortcut to all these calculations done in Figure 4 to Figure 7 can be reached by the usage of already established packages such as statsmodels (Seabold & Perktold, 2010), which includes the formular for an one-way ANOVA test (see Figure 8).

```
###One-way Anova using stat models###
import statsmodels.api as sm
from statsmodels.formula.api import ols

lm = ols('hospitalisation_value ~ vaccination_status',data=df_new).fit()
table = sm.stats.anova_lm(lm)
print(table)

                    df       sum_sq      mean_sq          F        PR(>F)
vaccination_status 2.0   544.445898   272.222949   51.542521   8.246473e-15
Residual          75.0   396.114134     5.281522         NaN            NaN
```

Figure 8: Code for the calculation of an one-way ANOVA test with the help of the python package "statsmodels" (Seabold & Perktold, 2010), on data set of hospitalized Corona patients with different vaccination status at the age of 18-59 (Robert-Koch Institut, 2022).

The critical F-value for this data set with degree of freedom in nominator of 2 and degree of freedom in denominator of 75 with an alpha of 0.05 would be 3.119 (see Supplementary table 3). The result of the F-statistics for Corona patients at the age of 18-59 have a F-value of 51.54. This means that there is a great variation between the sample means, indicating differences between the groups. To check if the differences are statistically significant, the p-value need to be analysed. The calculated p-value for hospitalized Corona patients with different vaccination status at the age of 18-59 is around 8.26e-15, which is less than the critical alpha value of 0.05. This means the rejection of the null hypothesis and that there is a significant difference between the three groups. To determine the specific differences a post-hoc test must be obtained. Therefore, the pairwise t-test was chosen as a post-hoc test with respect to the p-value adjustment according to the Bonferroni method. The code and result can be seen in Figure 9.

	Unvaccinated 18-59 years	Basic immunized 18-59 years	With booster vaccination 18-59 years
Unvaccinated 18-59 years	1.000000e+00	3.632689e-08	2.125473e-09
Basic immunized 18-59 years	3.632689e-08	1.000000e+00	8.964050e-05
With booster vaccination 18-59 years	2.125473e-09	8.964050e-05	1.000000e+00

Figure 9: Results of post-hoc pairwise T-Test (Seabold & Perktold, 2010) with Bonferroni p-value adjustment of data set from hospitalized Corona patients with different vaccination status at the age of 18-59 years.

Significant difference can be detected between the group of unimmunized patients and basic vaccinated patients, unimmunized patients, and patients with booster vaccination, as well as basic immunized and boosted patients.

Since the previously analysed data set includes only patients at the age of 18 to 59, the the results of the hospitalization of Corona patients with different vaccination status was tested, therefore the same code was used for the one-way ANOVA depicted in Figure 8.

```
lm = ols('Hospitalisierungswert ~ Impfstatus',data=df_new_60).fit()
table = sm.stats.anova_lm(lm)
print(table)

                df      sum_sq        mean_sq           F        PR(>F)
Impfstatus     2.0   12792.609869   6396.304934   70.458231   6.012866e-18
Residual      75.0    6808.613595     90.781515         NaN            NaN
```

Figure 10: Code for the calculation of an one-way ANOVA test with the help of the python package "statsmodels" (Seabold & Perktold, 2010), on data set of hospitalization of Corona patients with different vaccination status at the age of 60 plus (Robert-Koch Institut, 2022).

Since within this analysis, the same degree of freedoms applies, the critical F-value stays the same with 3.119 (see Supplementary table 3). The calculated F-value of the 60 plus aged patients is 70.46. This indicates a difference between the group. To check if the differences are statistically significant, the p-value needs to be analysed. The p-value of the 60 plus aged patients is 6.01e-18 and is less than 0.05. The differences in the group of this data set are significant. To identify the difference of the relationship between the groups, post-hoc analysis was carried out. The chosen post-hoc method was the pairwise t-test, because here all possibilities between the groups are tested.

The result of pairwise t-test as an post hoc test for the patient group aged 60 and plus showed, that a statistically significant difference between unvaccinated patients and basic immunized patients has been detected with an p-value of 9.578964e-10, also the hospitalization value of unvaccinated patients and patients with booster vaccination were calculated as statistically significant with the p-value of 1.625200e-11. Additionally, there is a significant difference between the hospitalization value of basic immunized and booster vaccinated patients (p-value 2.983115e-08). All three groups show significant differences between each other (see Figure 11).

	Unvaccinated 60+ years	Basic immunized 60+ years	With booster vaccination 60+ years
Unvaccinated 60+ years	1.000000e+00	9.578964e-10	1.625200e-11
Basic immunized 60+ years	9.578964e-10	1.000000e+00	2.983115e-08
With booster vaccination 60+ years	1.625200e-11	2.983115e-08	1.000000e+00

Figure 11: : Results of post-hoc pairwise T-Test (Seabold & Perktold, 2010) with Bonferroni p-value adjustment of data set from hospitalized Corona patients with different vaccination status at the age of 60 plus years.

With this results in hand, a conclusion is made in the following paragraph.

4. Conclusion

Parametric test are very powerful tools to identify differences in a mathematical proved way. The goal of this written assignment is to identify significant differences to hospitalization values of Corona patient with different vaccination status in Germany, therefore a mathematical method was used to detect if the variation of one independent variable (vaccination status) does have an effect to one dependent variable (hospitalization status). This issue was addressed by the one-way ANOVA method (see section 2.2). In total two one-way ANOVA tests were generated, since the data set was divided between a group of 18–59-year-old patient and a patient group at the age of 60 plus. Within both groups significant difference was detected with the one-way ANOVA test. The decisive reason was given by the calculated p-value (18-59 years: 8.26e-15, 60 plus years: 6.01e-18) which both were less than the set significance threshold of 0.05 (alpha). With the now mathematically proven difference the variating groups were of interest. To determine which groups, show significant differences, a post-hoc test weas conducted. Therefor the pairwise t-test was chosen, during this method multiple t-tests are performed in all possible comparisons are taken into consideration. Another advantage of the pairwise t-test in python, as it was used for this assignment (see Figure 9: and Figure 11:), is that the p-value adjustment method is included. In this case the Bonferroni method was chosen, which multiplies the p-value by the number of comparisons. The results show that all comparisons show significant differences. Meaning that independent of the age group, the hospitalization value of unvaccinated patients is significant different that basic immunized and boosted patients. Indicating that the vaccination does seem to have significant beneficial effects for the patients independent of the age, leading to less hospitalizations for these groups (see Figure 9: and Figure 11). Interestingly, also between the basic immunized patients and patients with booster vaccination was a significant difference detected for both age groups (age 18-59: 8.964050e-05, age 60 plus: 2.983115e-08). This leads to the conclusion that the course of disease in boosted patients is significantly weaker than in basic immunized patients independent of their age, so that fewer hospital admission was required. The beneficial of the vaccination does not only apply on the elderly patients at the age of 60 plus, but they also benefit significantly younger people at the age of 18 to 59 years. The visualization of the raw data (Robert-Koch Institut, 2022) does also reflect the effect of the vaccination on the hospitalization of Corona patients (see Figure 2 and Figure 3). All in all, it is now possible, after reviewing the data for patients in Germany from September 2021 to March 2022, to make the statement that not only the basic vaccination itself against corona but also the booster vaccination has a significant influence on the course of the disease. This influence seems to weaken the symptoms to the extent that hospitalisation is necessary in fewer and fewer cases for young and elderly people. Nevertheless, vaccinations themselves also carry a risk of side effects, which need to be taken into consideration individually.

Literature

Robert-Koch Institut. (2022, 28. April). *Inzidenzen der symptomatischen und hospitalisierten COVID-19-Fälle nach Impfstatus.* https://www.rki.de/DE/Content/InfAZ/N/Neuartiges_Coronavirus/Daten/Inzidenz_Impfstatus.html

Seabold, S. & Perktold, J. (2010). *9th Python in Science Conference: statsmodels: Econometric and statistical modeling with python.*

Thomas, K., Benjamin, R.-K., Fernando, P., Brian, G., Matthias, B., Jonathan, F., Kyle, K., Jessica, H., Jason, G., Sylvain, C., Paul, I., Dami'an, A., Safia, A. & Carol, W. (2016). *Jupyter Notebooks -- a publishing format for reproducible computational workflows: Positioning and Power in Academic Publishing: Players, Agents and Agendas*, S. 87–90.

World Health Organization. (2020). *WHO COVID-19 Dashboard.* https://covid19.who.int/

5. Appendix

Supplementary table 1: Hospitalised COVID-19 cases per 100,000, by vaccination status, aged 18-59 years (Robert-Koch Institut, 2022).

Meldejahr	Meldewoche	Ungeimpfte 18-59 Jahre	Grundimmunisierte 18-59 Jahre	Mit Auffrischimpfung 18-59 Jahre
2021	42	8	1,1	1,7
2021	43	9,4	1,3	2
2021	44	9,9	1,4	2
2021	45	12	1,7	0,5
2021	46	13,4	2,1	2
2021	47	14,2	2	1,5
2021	48	13,7	2	1,6
2021	49	12,3	2,1	1
2021	50	10,4	1,6	0,8
2021	51	7,6	1,3	0,5
2021	52	7,7	1,7	0,6
2022	1	6,4	2,3	0,8
2022	2	5	2,2	1
2022	3	5,6	2,2	0,9
2022	4	5,5	2,2	0,8
2022	5	6	1,9	0,8
2022	6	5,4	2	0,9
2022	7	4,6	1,6	0,8
2022	8	3,9	1,5	0,9
2022	9	3,4	1,7	1
2022	10	3,1	1,6	0,9
2022	11	3,3	1,3	0,9
2022	12	2,9	1,3	0,8
2022	13	2,4	1,1	0,8
2022	14	1,9	0,7	0,6
2022	15	1,4	0,5	0,5

Supplementary table 2: Hospitalised COVID-19 cases per 100,000, by vaccination status, aged 60 plus years (Robert-Koch Institut, 2022).

Reporting year	Reporting week	Unvaccinated 60 plus years	Basic immunized 60 plus years	With booster vaccination 60 plus years

2021	42	28,1	5,8	4,0
2021	43	32,3	7,5	5,1
2021	44	38,5	8,5	5,5
2021	45	48,6	10,2	6,0
2021	46	55,8	10,9	7,4
2021	47	63,4	11,2	6,4
2021	48	62,2	9,9	5,2
2021	49	64,7	9,8	4,9
2021	50	52,8	8,1	3,2
2021	51	38,5	5,2	1,7
2021	52	37,6	5,6	1,8
2022	1	29,5	5,5	1,6
2022	2	20,3	4,1	1,7
2022	3	16,5	5,2	2,1
2022	4	20,8	6,0	2,0
2022	5	24,6	7,3	2,5
2022	6	27,7	7,4	2,6
2022	7	26,2	6,9	2,9
2022	8	23,3	7,2	3,1
2022	9	21,6	6,0	3,0
2022	10	20,6	6,1	3,3
2022	11	21,8	8,1	3,2
2022	12	21,7	6,2	2,7
2022	13	16,5	5,4	2,4
2022	14	13,4	4,4	1,9
2022	15	8,9	3,3	1,4

Supplementary table 3: Critical values for "F" Distribution with alpha as 0.05. Critical "F" value marked in red square for degree of freedoms numerator 2 and denominator of 75.

CRITICAL VALUES for the "F" Distribution, ALPHA = .05.

Denominator DF	Numerator DF									
	1	2	3	4	5	6	7	8	9	10
51	4.030	3.179	2.786	2.553	2.397	2.283	2.195	2.126	2.069	2.022
52	4.027	3.175	2.783	2.550	2.393	2.279	2.192	2.122	2.066	2.018
53	4.023	3.172	2.779	2.546	2.389	2.275	2.188	2.119	2.062	2.015
54	4.020	3.168	2.776	2.543	2.386	2.272	2.185	2.115	2.059	2.011
55	4.016	3.165	2.773	2.540	2.383	2.269	2.181	2.112	2.055	2.008
56	4.013	3.162	2.769	2.537	2.380	2.266	2.178	2.109	2.052	2.005
57	4.010	3.159	2.766	2.534	2.377	2.263	2.175	2.106	2.049	2.001
58	4.007	3.156	2.764	2.531	2.374	2.260	2.172	2.103	2.046	1.998
59	4.004	3.153	2.761	2.528	2.371	2.257	2.169	2.100	2.043	1.995
60	4.001	3.150	2.758	2.525	2.368	2.254	2.167	2.097	2.040	1.993
61	3.998	3.148	2.755	2.523	2.366	2.251	2.164	2.094	2.037	1.990
62	3.996	3.145	2.753	2.520	2.363	2.249	2.161	2.092	2.035	1.987
63	3.993	3.143	2.751	2.518	2.361	2.246	2.159	2.089	2.032	1.985
64	3.991	3.140	2.748	2.515	2.358	2.244	2.156	2.087	2.030	1.982
65	3.989	3.138	2.746	2.513	2.356	2.242	2.154	2.084	2.027	1.980
66	3.986	3.136	2.744	2.511	2.354	2.239	2.152	2.082	2.025	1.977
67	3.984	3.134	2.742	2.509	2.352	2.237	2.150	2.080	2.023	1.975
68	3.982	3.132	2.740	2.507	2.350	2.235	2.148	2.078	2.021	1.973
69	3.980	3.130	2.737	2.505	2.348	2.233	2.145	2.076	2.019	1.971
70	3.978	3.128	2.736	2.503	2.346	2.231	2.143	2.074	2.017	1.969
71	3.976	3.126	2.734	2.501	2.344	2.229	2.142	2.072	2.015	1.967
72	3.974	3.124	2.732	2.499	2.342	2.227	2.140	2.070	2.013	1.965
73	3.972	3.122	2.730	2.497	2.340	2.226	2.138	2.068	2.011	1.963
74	3.970	3.120	2.728	2.495	2.338	2.224	2.136	2.066	2.009	1.961
75	3.968	3.119	2.727	2.494	2.337	2.222	2.134	2.064	2.007	1.959
76	3.967	3.117	2.725	2.492	2.335	2.220	2.133	2.063	2.006	1.958
77	3.965	3.115	2.723	2.490	2.333	2.219	2.131	2.061	2.004	1.956
78	3.963	3.114	2.722	2.489	2.332	2.217	2.129	2.059	2.002	1.954
79	3.962	3.112	2.720	2.487	2.330	2.216	2.128	2.058	2.001	1.953
80	3.960	3.111	2.719	2.486	2.329	2.214	2.126	2.056	1.999	1.951